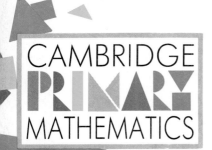

CAMBRIDGE PRIMARY MATHEMATICS

MODULE I
BOOK 2

Roy Edwards
Mary Edwards
Alan Ward

Cambridge University Press
Cambridge
New York   Port Chester
Melbourne   Sydney

Published by the Press Syndicate of the University of Cambridge
The Pitt Building, Trumpington Street, Cambridge CB2 1RP
40 West 20th Street, New York, NY 10011–4211, USA
10 Stamford Road, Oakleigh, Victoria 3166, Australia

First published 1989
Third printing 1991

Printed in Great Britain by Scotprint Ltd, Musselburgh

British Library cataloguing in publication data

Edwards, Roy
Cambridge primary mathematics.
Module 5
Bk. 2
1. Mathematics
I. Title   II. Edwards, Mary   III. Ward, Alan
510

ISBN 0 521 35823 X

The authors and publishers would like to thank the many schools
and individuals who have commented on draft material for this
course. In particular, they would like to thank Anita Straker for
her contribution to the suggestions for work with computers,
Norma Anderson, Ronalyn Hargreaves (Hyndburn Ethnic
Minority Support Service) and John Hyland Advisory Teacher
in Thameside, and the staff and pupils of St Laurence School,
Cambridge, St Mary's Junior School, Ely, Riverside Middle School,
Mildenhall, Teversham C of E (A) School.

Photographs are reproduced courtesy of:
front cover, pp 6, 76 ZEFA; pp 11, 12 Sefton Photo Library;
pp 26–7 Spanish National Tourist Office; p 34 British Airports
Services; pp 36-7 Quadrant Picture Library; p 65 A.N.T./NHPA,
John Shaw/NHPA, R. Knightbridge/NHPA, Douglas Dickins/NHPA,
Melvin Grey/NHPA, Anthony Bannister/NHPA;
p 66 Peter Johnson/NHPA, Stephen Dalton NHPA;
p 68 Peter Pickford/NHPA. All other photographs by Graham Portlock,

The mathematical apparatus was kindly supplied by E J Arnold,
the balance (p 69) by F H Fry and the calendar (p 77) by the
Amaising Publishing House.

Designed by Chris McLeod

Illustrations by Chris Ryley
Diagrams by DP Press
Children's graphs by George McLeod

DP

Contents

Data 2

A

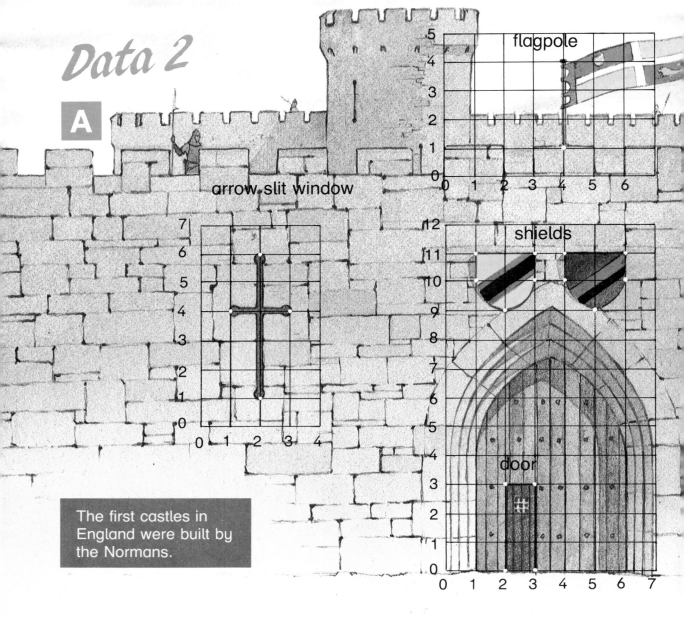

flagpole

arrow slit window

shields

door

The first castles in England were built by the Normans.

The co-ordinates of the bottom of the flagpole are (4,1).
Write the co-ordinates of these.

1. the top of the flagpole (☐,☐)

2. the window (☐,☐) (☐,☐) (☐,☐) (☐,☐)

3. the shields (☐,☐) (☐,☐) (☐,☐) (☐,☐) (☐,☐)
   (☐,☐) (☐,☐) (☐,☐) (☐,☐) (☐,☐)

4. the small door (☐,☐) (☐,☐) (☐,☐) (☐,☐)

4

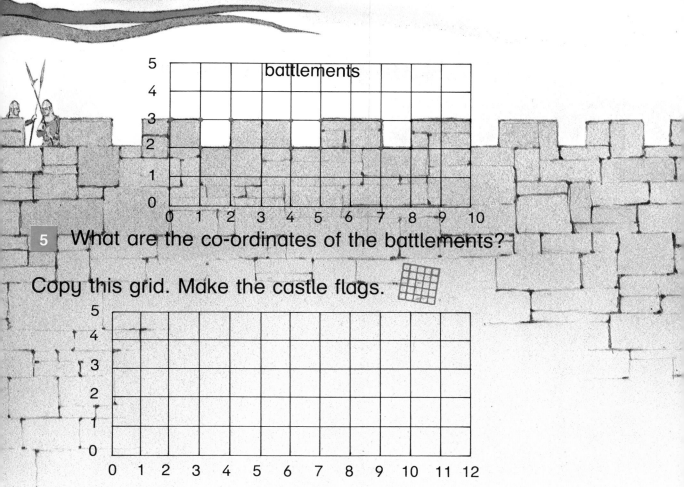

5 What are the co-ordinates of the battlements?

Copy this grid. Make the castle flags.

Join the co-ordinates in order to make three flags.

6 (1,1) (1,4) (4,1) (1,1)

7 (5,1) (5,3) (8,3) (8,1) (5,1)

8 (9,1) (9,4) (12,4) (12,1) (9,1)

9 What shapes have you made?

Let's investigate

Copy the grid again.
Plan co-ordinates for a castle key.
Write them in order. Draw it.

This is the plan of a castle.

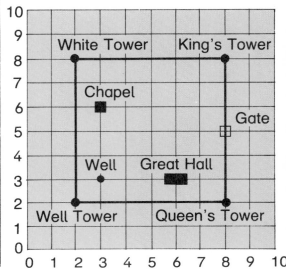

Write the co-ordinates of these.

1. the Well Tower

2. the Queen's Tower

3. the King's Tower

4. the White Tower

5. the Great Hall

6. the chapel

7. the well

8. the gate

9. Copy the castle plan.
   Join these co-ordinates to make an outer wall.
   (1,1)   (1,9)   (9,9)   (9,1)   (1,1)

10. Draw these on your plan.
    Write their co-ordinates.

    | Outer Gate |
    | Stables |
    | Kitchen |
    | Store |

11. Plan some towers in the outer wall. Write their co-ordinates.

Knights wore armour in battle. They recognised each other by the patterns or pictures on their shields. These were called coats of arms.

Copy these shields.
Join the co-ordinates to make coats of arms.

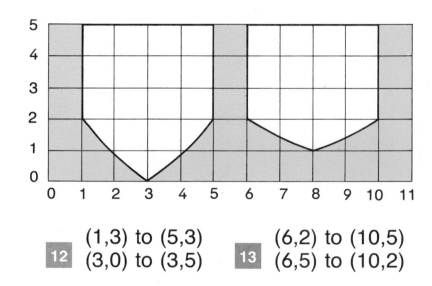

12 (1,3) to (5,3)
(3,0) to (3,5)

13 (6,2) to (10,5)
(6,5) to (10,2)

Let's investigate

Make up some more coats of arms.
Plan the patterns and write the co-ordinates.

C Let's investigate

Make up your own plan for a castle.
Put on towers, buildings and a gate.
Draw a wall round the outside.
Write the co-ordinates of the things you have drawn.

1 Four programmes cost ☐ p. 4 × ☐ p = ☐ p

2 Six programmes cost ☐ p.

3 Ten programmes cost ☐ p.

45p = £0·45

4 Write each answer in pounds.

Find the cost of these tickets in two ways.

5 3 tickets for Thursday ☐ p + ☐ p + ☐ p = £ ☐

or 3 × ☐ p = £ ☐

6 4 tickets for Saturday.　　**7** 4 tickets for Friday.

8 5 tickets for Thursday.

Let's investigate

Show different ways to pay for a 30p ticket.
Use sets of the same coin each time.

☐ × 1 p = ☐ p　　☐ × 2 p = ☐ p

B

1 What is the total cost of 3 tickets and 3 programmes for Thursday?

2 What is the total cost of 2 tickets and 1 programme for Saturday?

How much change is there from 90p?

3 What is the total cost of 1 ticket for each night?

How much change is there from £1?

4 How many tickets can be bought for 80p each night?

9

Refreshments

Tea 12p   Orange drink 11p
Buns 18p   Cakes 25p

Tickets

	Adults	children
Thursday	25p	12p
Friday	30p	15p
Saturday	35p	18p

Find the cost of these.

5 7 buns and 2 orange drinks.

6 3 cakes and 3 teas.

7 3 orange drinks and 2 teas.

8 5 teas and 2 buns.

Let's investigate

Make a chart to show the cost of cups of tea. Stop at 20 cups.

C Find the cost of these.

1 3 adults and 1 child for Saturday.

2 3 children and their mother on Friday.

3 2 adults and 1 child on Thursday.

4 4 adults and 4 children on Thursday.

Let's investigate

25 children went to the school play on Saturday.
How much did they pay altogether?
Try to work it out without using a calculator.
Explain how you did it.

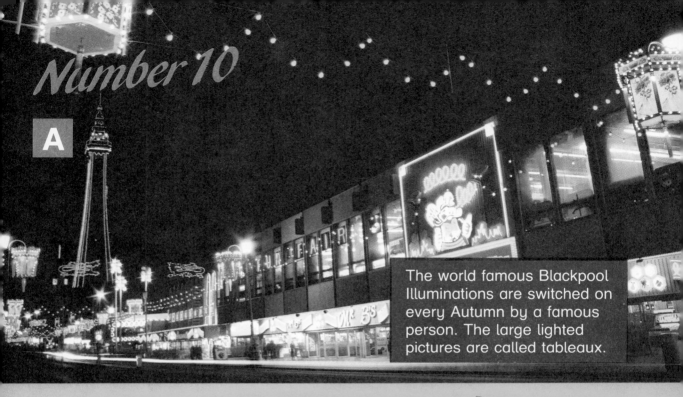

A

The world famous Blackpool Illuminations are switched on every Autumn by a famous person. The large lighted pictures are called tableaux.

Plan the coloured lighting for these tableaux.

1 Colour $\frac{1}{3}$ yellow, $\frac{1}{3}$ red, $\frac{1}{3}$ green.
How many thirds are coloured?

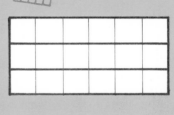

2 Colour $\frac{1}{3}$ yellow and $\frac{2}{3}$ green.
Do it again in a different way.

3 Colour $\frac{1}{3}$ red and $\frac{1}{3}$ blue.
How much is left?
Do it again in a different way.

Let's investigate

Colour $\frac{1}{3}$ red, $\frac{1}{3}$ green and $\frac{1}{3}$ yellow.
Find other ways to do it.

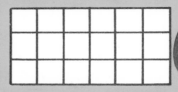

Count the squares

Once the illuminations are over it takes about 6 weeks to take them down. The thousands of bulbs are tested, washed and resprayed ready for next year.

1. $\frac{\square}{3}$ of the whole box have bulbs in.

2. $\frac{\square}{3}$ of the whole box is empty.

3. How many bulbs are there altogether?

4. $\frac{\square}{3}$ are in the first part of the box.

5. $\frac{\square}{3}$ are in the last part.

What fraction of all the bulbs are these?

6. The purple bulbs

7. The yellow bulbs

8. The large bulbs

9. The small bulbs

10. How many bulbs would you move to the last part of the box so that $\frac{2}{3}$ of the bulbs are there?

Copy the shapes of these lights.
Complete them to show the whole shapes.

11

$\frac{1}{3}$

12

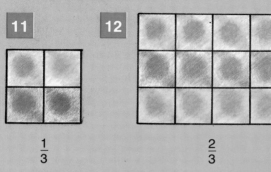

$\frac{2}{3}$

13

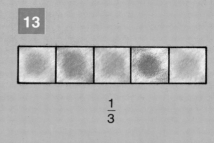

$\frac{1}{3}$

Let's investigate

Draw 2 boxes of light bulbs. Choose the number of bulbs.
Put $\frac{1}{3}$ in one box and $\frac{2}{3}$ in the other.
Do the same for other numbers of bulbs.

C Draw the whole shapes for these lights.

1

$\frac{1}{3}$

$\frac{1}{3}$

2

$\frac{1}{3}$

3

$\frac{1}{3}$

4

$\frac{1}{3}$

5

$\frac{1}{3}$

Let's investigate

Stamp some clock faces in your book.
Use them to show different ways of dividing
a circle into thirds. Colour them.

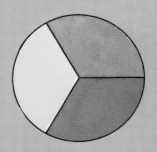

Length 2

Perimeter is the distance all the way round the shape.

A

Use a ruler.
Find the perimeter of these shapes.
☐ cm + ☐ cm + ☐ cm + ☐ cm.

1

2

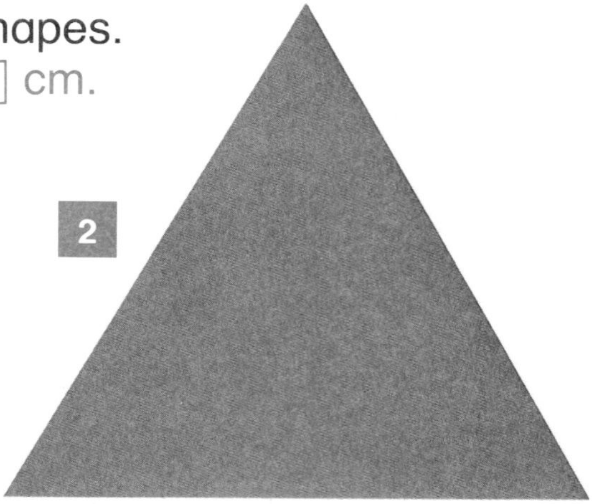

3

4

5　What is the largest perimeter?

6　What is the smallest perimeter?

7　What is the difference between the largest and smallest perimeters?

8　What is the difference between the perimeters of the square and the rectangle?

14

9 Find the perimeter of the yellow rectangle.

10 Find the perimeter of the blue rectangle.

Let's investigate

Draw different shapes with perimeters of 24 cm.

B Find the perimeters of these shapes.

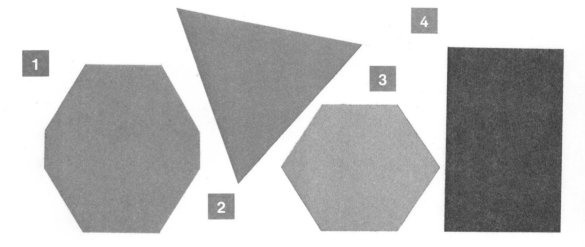

5 Which shape has the largest perimeter?

6 What is the difference between the perimeters of the triangle and rectangle?

15

7 Each side of this squared board is 40 cm.
Find the perimeter.

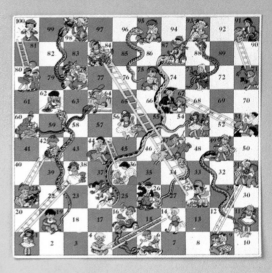

8 Find the perimeter of each small square.
Show how you did it.

9 Double the size of each small square.
Find the perimeter of the board now.

Let's investigate

Make different shapes each with a perimeter of 14 cm.
Use six squares with 1 cm sides each time.

C **1** Use six equilateral triangles of the same size to make a shape. The perimeter must be 6 times longer than the side of a triangle.

Let's investigate

Use 9 squares.
Put them together to make shapes.
What different perimeters can you make?
What is the largest? What is the smallest?

Weight 2

soldier
10g red
10g black
10g white

tiger
20g brown
20g yellow

panda
25g black
25g white

A How many grams of wool for these toys?

1 one soldier ☐ g

2 one tiger ☐ g

3 one panda ☐ g

4 three soldiers ☐ g

5 two tigers ☐ g

6 two pandas ☐ g

7 Two chicks weigh 18g.
One chick weighs ☐ g.

8 Three teddies weigh 60 g.
One teddy weighs ☐ g.

Let's investigate

Weigh some soft toys.
Why are they different weights?
How could you make them heavier?

B

53g red
53g dark red

92g green
92g pale green

40g blue
40g dark blue
40g pale blue

26g yellow
26g pale yellow

How much do these weigh?

1 the blue mittens

2 the red mittens

3 the yellow mittens

4 the green mittens.

5 How much dark blue wool for 4 pairs of blue mittens?

6 How much dark red wool for 3 pairs of red mittens?

7 A pair of socks weighs 88 g. One sock will weigh ☐ g.

8 A pair of gloves weighs 76 g. One glove will weigh ☐ g.

Let's investigate

Weigh a jumper.
Find some other clothes that you think weigh about the same
as the jumper. Weigh them to see if you were right.

C

1. What is the heaviest weight of wool for the scarf?

2. What is the lightest weight of wool for the hat?

3. What is the most the scarf, hat and gloves could weigh together?

4. What is the lightest weight of wool you would need to make them all?

Let's investigate

Each red ball of wool weighs 25 g. Each yellow ball weighs 40 g.

Find ways to make up 300 g, 400 g and 500 g of red and yellow wool.

gloves
three 25g balls or two 40g balls

hat
two 40g balls or three 25g balls or two 50g balls

scarf
three 40g balls or five 25g balls or two 50g balls

capacity 2

A

1 The bottle of orange juice holds ☐ ml.

2 How many glasses will it fill?

3 The carton holds ☐ ml.

4 How many beakers will it fill?

A medicine spoon holds 5 ml.

5 2 spoons hold ☐ ml.   **6** 4 spoons hold ☐ ml.

7 10 spoons hold ☐ ml.

8 The medicine bottle holds ☐ spoonfuls.

How long will the medicine last?

9 | 2 spoonfuls each day | ☐ days.   **10** | 4 spoonfuls each day | ☐ days.

beaker
200 ml

medicine
100 ml

glass
250 ml

11 Grandad drinks four cups of tea. This is ☐ ml.

12 Jim drinks three mugs of orange. This is ☐ ml.

13 Who drinks most?

14 He drinks ☐ ml more.

Let's investigate

Find things that hold less than 25 ml.
Make a list of them.
Measure how much each one holds.

B

1 The kettle holds ☐ ml.

2 How many full cups will the teapot fill?

3 How much is left in the teapot now?

4 How many mugs will the teapot fill?

5 How much is left in the teapot?

6 A small cup holds 150 ml.
How many will the teapot fill?

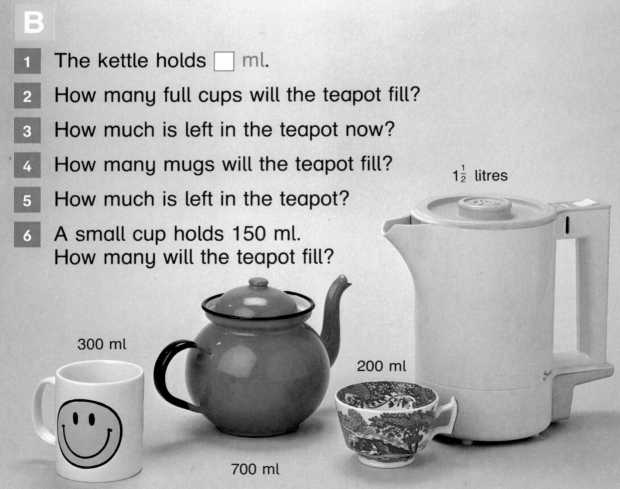

1½ litres

300 ml

200 ml

700 ml

21

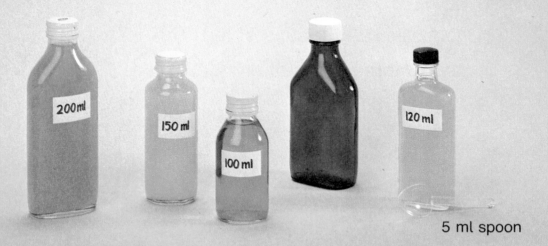

5 ml spoon

Which bottle fits which label? The spoon holds 5 ml.

7 | 3 spoonfuls twice a day for 5 days.

8 | 2 spoonfuls twice a day for 5 days.

9 | 1 spoonful four times a day for 10 days.

10 How much will be in the brown bottle?

11 How much is left after 3 days?

2 spoonfuls three times a day for 6 days.

Let's investigate

Write different labels for the green medicine.
Try to use it all up.

C **1** Find the weight of a litre of water.
How did you do it?

Let's investigate

Use only a large plastic bottle, funnel and scales.

You must put exactly half a litre of water into the bottle.
How will you do it?

Time 2

10 a.m.   1 p.m.   4 p.m.

1. The school trip started at
8 o'clock in the morning.
Are the times on the pictures
before or after midday?

a.m. times are from
midnight to midday,
or noon.
p.m. times are from
midday to midnight.

Are these times before or after midday?

2. 9 a.m.   3. 3 p.m.   4. 7 p.m.   5. 11 a.m.

6. Draw things that
you do at these times.

9 a.m.   11 a.m.   2 p.m.   7 p.m.

7. Draw 2 things that
always happen a.m.
Draw 2 things that
always happen p.m.

a.m.   a.m.
p.m.   p.m.

8. Write these times in order. Start with the earliest.
11 a.m.   2 p.m.   6 a.m.   5 p.m.

Write a.m. or p.m. times.

Let's investigate

Think of times when it would be amusing if a.m. and p.m. times were changed.

B

Write these times putting a.m. or p.m.

1 half an hour after noon

2 half an hour after midnight

Copy the time line.
Write times for these. Draw arrows to show them.

3 getting up   **4** morning play time

5 lunch   **6** going to bed

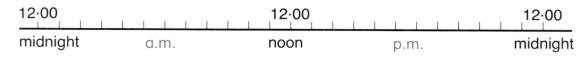

12·00   12·00   12·00

midnight   a.m.   noon   p.m.   midnight

7 Put these times in order on a time line.
Start at 10·00 a.m.

| 10·00 a.m. | 9·00 p.m. | 2·00 a.m. | 11·00 a.m. | 10·00 p.m. |

How many hours between these times?

8 7·00 a.m. to 2·00 p.m.   **9** 11·00 a.m. to 5·00 p.m.

10 11·00 p.m. to 5·00 a.m.   **11** 7·00 p.m. to 3·00 a.m.

Let's investigate

This is a time line for the school trip.

8·00 a.m.   12·00   6·00 p.m.

start   noon   home

Copy the line.
Show when you might find busy traffic.
Say why you think this is.

C *Let's investigate*

Plan a school trip.
Draw the things you
would do and see.
Write times for
each one.
Don't forget to eat.

A Every year thousands of people fly to sunny countries for their summer holidays.

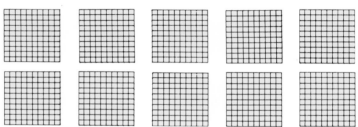

10 hundreds   = 1 thousand or 1000

14 hundreds   = 1400

1   16 hundreds = ☐   **2**   19 hundreds = ☐

3   13 hundreds = ☐   **4**   15 hundreds = ☐

5   1200 = ☐ hundreds   **6**   1800 = ☐ hundreds

Copy and finish these number patterns.

7   1000   1100   1200   1300   ☐ ☐ ☐ ☐ ☐ ☐   2000

8   1000   1200   ☐ ☐   1800   ☐

9   1800   1700   ☐ ☐ ☐ ☐ ☐ ☐   1000

10 What number is on each colour?

grey 1225   yellow ☐

blue ☐   green ☐

11 Write the numbers on
the colours in order.
Start with the smallest.

12 How many altogether are
on the grey and yellow?

13 How many altogether are
on the blue and green?

14 How many more are on
the green than on
the grey?

15 How many more are on
the yellow than on
the blue?

Let's investigate

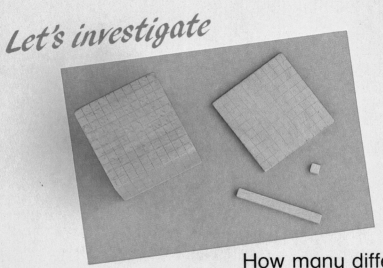

How many different numbers
can you make using any of these?

B

1 Which pair of numbers
add up to 7479?

2 Which pair adds
up to 7660?

3 Write four thousand six hundred
and seventy in numbers.

3242 *3107*
4263 *3517*
4418 *4372*

Look at the picture of the plane.
Write the distances between these places.

4 London and San Francisco. **5** Nairobi and London.

6 Moscow and Tokyo. **7** Bombay and London.

8 How much further is it from London to Hong Kong
than from London to Tokyo?

9 Write the distances shown in order.
Start with the longest.

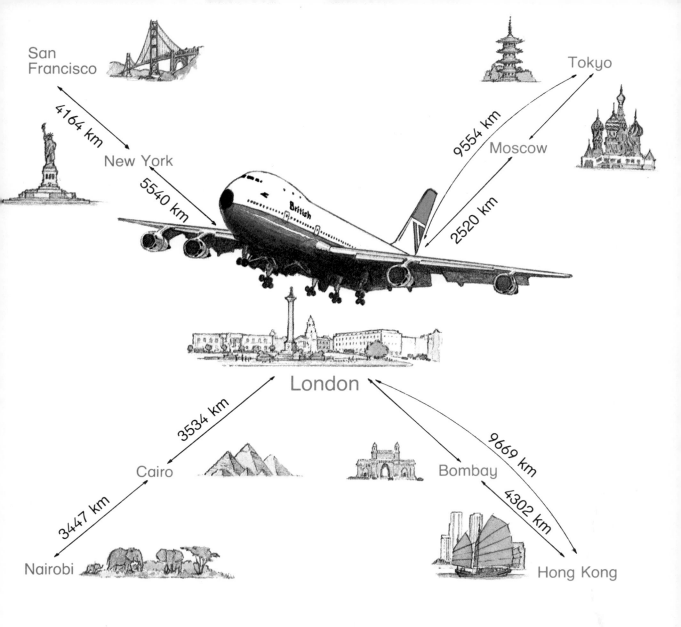

San Francisco — 4164 km — New York

New York — 5540 km — London

Tokyo — 9554 km — Moscow

Moscow — 2520 km — London

London — 3534 km — Cairo

Cairo — 3447 km — Nairobi

London — 9669 km — Hong Kong

Bombay — 4302 km — Hong Kong

Let's investigate

Plan some distances for these journeys.

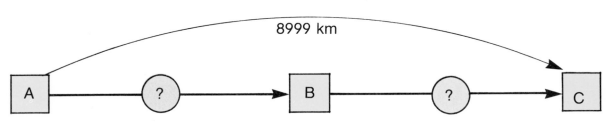

8999 km

A → ? → B → ? → C

C

	Rome	Paris	Ottawa	New York	Madrid	Lisbon	Geneva	Athens
Athens	1043	2103	7760	7923	2360	2853	1719	▇
Geneva	715	394	6044	6202	1009	1496	▇	1719
Lisbon	1869	1439	5385	5406	514	▇	1496	2853
Madrid	1359	1032	5701	5764	▇	514	1009	2360
New York	6892	5830	528	▇	5764	5406	6202	7923
Ottawa	6723	5664	▇	528	5701	5385	6044	7760
Paris	1109	▇	5664	5830	1032	1439	394	2103
Rome	▇	1109	6723	6892	1359	1869	715	1043

This chart shows distances in kilometres between the cities.
Find the places on a map.

Find the distances between these cities.

1 New York and Paris.
2 Ottawa and Lisbon.
3 Madrid and Athens.
4 Lisbon and Rome.
5 Geneva and Ottawa.
6 Rome and New York.
7 Which two cities are the furthest apart?
8 Which two cities are nearest to each other?

Let's investigate

Use the chart to plan some journeys.
Each journey must be less than ten thousand km.

| start ? | → | | → | | → | | = | | km |

Angles 2

A

1 right-angle
is 90°

$\frac{1}{2}$ right-angle
is 45°

1 How many $\frac{1}{2}$ right-angles can you see?

What is the direction from the teacher to each of these things?

2 the board

3 the computer

4 the window

How many $\frac{1}{2}$ right-angles between these? Go the shortest way.

5 NE and SE

6 N and S

7 SW and E

How many degrees between these? Go the shortest way.

8 N and W

9 S and SE

10 NW and NE

11 door and window.

12 board and sink.

Let's investigate

Use a compass.
Find the directions of places in school from your classroom.

SCOTLAND

Aberdeen

Perth

Edinburgh

Glasgow

Alnwick

Dumfries

Belfast

IRELAND

York

Blackpool

Leeds

Bradford

Grimsby

Dublin

Sheffield

Lincoln

Llandudno

Caernarfon

Chester

ENGLAND

Aberystwyth

Coventry

Worcester

WALES

Oxford

Cardiff

Swindon

Bristol

Reading

London

Southampton

Brighton

Penzance

B

1   Use your compass to finish this table.

London	→ Grimsby	North	Aberdeen	→ Glasgow	_____
Bradford	→ Swindon	_____	Worcester	→ Lincoln	_____
Llandudno	→ Belfast	_____	Bristol	→ Reading	_____
Sheffield	→ Dublin	_____	Blackpool	→ Caernarfon	_____
Dumfries	→ Leeds	_____	Oxford	→ York	_____

2   Follow this trail.

Start at Alnwick.
Go NW to a city in Scotland. It starts with P.
Go S to a city in Wales. It starts with C.
Where are you?

Let's investigate

Use the map to make up your own trail.
Give three directions.
Let a friend follow it.

C

1   Use the map.
London is _____ from Brighton.
London is _____ from Southampton.
London is _____ from Chester.
London is _____ from Cardiff.

Let's investigate

Choose another town or city on the map.
Give its directions from other places.

Number 12

London
286 people get on

Toronto
150 people get off

Vancouver
everyone gets off

1 75 children get on the plane at London.
How many adults get on?

2 45 children get off the plane at Toronto.
How many adults get off?

3 How many people get off at Vancouver?

4
$$\begin{array}{r} 4\ 2\ 7 \\ -\ 1\ 6\ 5 \\ \hline \end{array}$$

5
$$\begin{array}{r} 5\ 4\ 6 \\ -\ 3\ 7\ 2 \\ \hline \end{array}$$

6
$$\begin{array}{r} 6\ 3\ 8 \\ -\ 2\ 5\ 3 \\ \hline \end{array}$$

7
$$\begin{array}{r} 6\ 3\ 0 \\ -\ 4\ 6\ 7 \\ \hline \end{array}$$

8
$$\begin{array}{r} 5\ 2\ 3 \\ -\ 2\ 5\ 9 \\ \hline \end{array}$$

9
$$\begin{array}{r} 7\ 2\ 2 \\ -\ 1\ 8\ 5 \\ \hline \end{array}$$

Use the numbers on the pink and orange labels.

10 Add 100 to each.   11 Add 99 to each.

12 Take 100 from each.   13 Take 99 from each.

Use the yellow and green labels.

14 Write the numbers in order. Start with the largest.

15 Take 1000 from each.

16 Write each number to the nearest 1000.

17 Which numbers are even?

Let's investigate

Show 8 on your calculator.
Keep subtracting 2.

$$8 \rightarrow \square \overset{-2}{\rightarrow} \square \overset{-2}{\rightarrow} \square \overset{-2}{\rightarrow} \square \overset{-2}{\rightarrow} \square \overset{-2}{\rightarrow} \square$$

What happens after zero?
Make up some more patterns that go below zero.

B

1 314 people boarded a plane.
86 were children.
How many were adults?

2 182 passengers were men and boys.
How many were women and girls?

These puzzles were in the flight magazine.
Copy and complete them.

3

250	+	127	=	
+	■	+	■	+
196	+	148	=	
=	■	=	■	=
	+		=	

4

721	−	344	=	
−	■	−	■	−
275	−	148	=	
=	■	=	■	=
	−		=	

Complete these patterns.

5   1350   1400   ☐   ☐   ☐   ☐   ☐   1700

6   1000   1250   ☐   ☐   ☐   ☐   ☐   2750

7   2300   2200   2100   ☐   ☐   ☐   1700

8   1950   1750   ☐   ☐   ☐   950

36

9 Write each number to the nearest thousand.

10 Subtract 1000 from each number.

11 Subtract 999 from each number.

12 Add 1000 to each number.

13 Add 999 to each number.

Find the mystery numbers.

14 I am odd.
My tens digit is even.
I am between 348 and 372.

15 I am even.
My digits add up to 9.
My hundreds digit is odd.
I am between 1145 and 1432.

37

Let's investigate

Choose some numbers.
Pick one to be the mystery number.
Write some clues for it.
Ask a friend to find it.

C

1 Copy this shape.

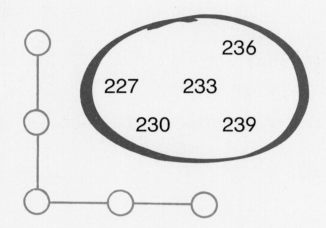

227   233   236
230   239

Put one of these numbers in each circle.
Use each number only once.
The numbers going down must add up to
the same as the numbers going across.

Let's investigate

Make up four different numbers for the empty circles.
The numbers going down must add up to the same
as the numbers going across.

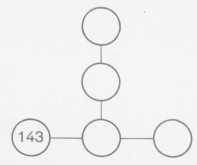

Find other ways to do it.

38

Shape 3

A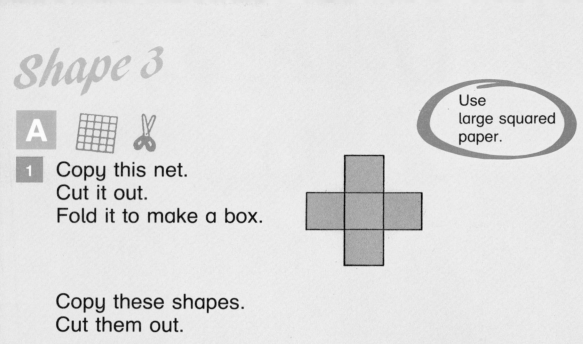

1 Copy this net.
Cut it out.
Fold it to make a box.

Use
large squared
paper.

Copy these shapes.
Cut them out.

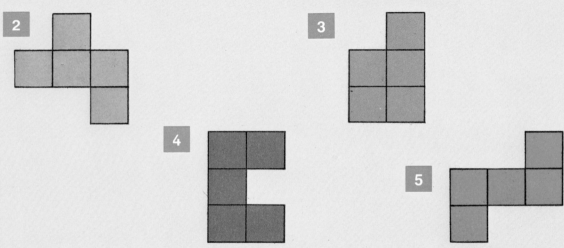

2

3

4

5

6 Which ones make a box?

7 What shape are the boxes?

Let's investigate

Draw some other shapes with five squares.
How many different nets can you find that
will make a box?

B

1 Some boxes have lids.
Copy this net onto
paper with large squares.
Cut it out.
What shape can you make?

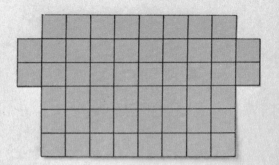

Copy these shapes.
Cut them out.

2

3

4

5

6 Which will fold to make boxes with lids?
Make two of them.

7 Use one of the boxes to make a die.
Use the numbers 1 to 6.
Draw a number on each face.
Opposite faces must add up to 7.

Let's investigate

Open out a box.
Draw its net on squared paper.
Ask a friend to make it from the net.
Try this with other boxes.

C

1 Use an equilateral triangle template.
Make a net like this. Cut it out.
What box shape can you make?

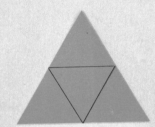

2 Find a way to make
the same box using
this net.

Hint: Tuck in the last triangle.

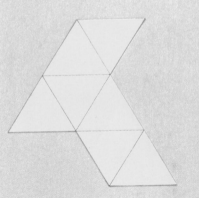

Let's investigate

Draw the nets of two boxes.
One must fit inside the other.
Make the boxes. Can you make a set of three boxes?

41

X	1	2	3	4	5	6	7	8	9	10
1	1	2	3	4	5	6	7	8	9	10
2	2	4	6	8	10	12	14	16	18	20
3	3	6	9	12	15	18	21	24	27	30
4	4	8	12	16	20	24	28	32	36	40
5	5	10	15	20	25	30	35	40	45	50
6	6	12	18	24	30	36	42	48	54	60
7	7	14	21	28	35	42	49	56	63	70
8	8	16	24	32	40	48	56	64	72	80
9	9	18	27	36	45	54	63	72	81	90
10	10	20	30	40	50	60	70	80	90	100

$3 \times 4 = 12$
3 and 4 are
factors of 12.

Use the multiplication square to do these.

1 $6 \times \square = 18$ 2 $2 \times \square = 12$ 3 $8 \times \square = 64$

4 $\square \times 10 = 10$ 5 $\square \times 2 = 18$ 6 $\square \times 5 = 15$

7 $5 \times \square = 20$ 8 $\square \times 9 = 63$ 9 $\square \times 7 = 28$

10 $7 \times \square = 21$ 11 $6 \times \square = 36$ 12 $\square \times 1 = 5$

Find factors to fit the boxes.

13 ☐ × ☐ = 35   **14** ☐ × ☐ = 12   **15** ☐ × ☐ = 64

16 ☐ × ☐ = 10   **17** ☐ × ☐ = 24   **18** ☐ × ☐ = 15

19 ☐ × ☐ = 30   **20** ☐ × ☐ = 63   **21** ☐ × ☐ = 28

22
$$\begin{array}{r} 1\ 7 \\ \times\ \underline{\quad 8} \\ \hline \end{array}$$

23
$$\begin{array}{r} 5\ 2 \\ \times\ \underline{\quad 7} \\ \hline \end{array}$$

24
$$\begin{array}{r} 4\ 6 \\ \times\ \underline{\quad 6} \\ \hline \end{array}$$

25
$$\begin{array}{r} 4\ 3 \\ \times\ \underline{\quad 9} \\ \hline \end{array}$$

26 Use a multiplication square.
Colour the table of 1 in blue.
Start at 1 × 1 = 1   1 × 1 = 1
         2 × 1 = 2   1 × 2 = 2
Colour as far as
       10 × 1 = 10   1 × 10 = 10

X	1	2	3
1			
2			
3			

27 Do the same for the table of 10.
        1 × 10 = 10   10 × 1 = 10
        2 × 10 = 20   10 × 2 = 20
Colour as far as 10 × 10.
What do you notice?

Let's investigate

Use a multiplication square.
Colour the tables of 2, 4 and 8
in different colours.
What do you notice?

B Use multiplication squares.

X	1	2	3	4	5	6	7	8	9	10
1	1	2	3	4	5	6	7	8	9	10
2	2	4	6	8	10	12	14	16	18	20
3	3	6	9	12	15	18	21	24	27	30
4	4	8	12	16	20	24	28	32	36	40
5	5	10	15	20	25	30	35	40	45	50
6	6	12	18	24	30	36	42	48	54	60
7	7	14	21	28	35	42	49	56	63	70
8	8	16	24	32	40	48	56	64	72	80
9	9	18	27	36	45	54	63	72	81	90
10	10	20	30	40	50	60	70	80	90	100

1 Colour all the numbers that have a factor of 5 in red.

Colour all the numbers that have a factor of 10 in blue.

What do you notice?

2 Write the numbers that have factors of both 5 and 10.

3 6 has four factors. Use the multiplication square to find them.

X	1	2	3	4	5	6
1	1	2	3	4	5	⑥
2	2	4	⑥	8	10	12
3	3	⑥	9	12	15	18

4 What are the factors of 8?

5 What are the factors of 9?

6 Copy this diagram. Put the factors of 12 in the circles. Opposite numbers should multiply to give 12.

7 Do the same for 15.

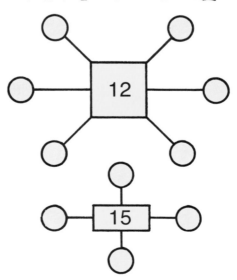

44

Let's investigate

Use a multiplication square.
Find numbers that have 2 pairs of factors.

Now find ones that have 3 pairs of factors.

Find numbers with more pairs of factors.

Cover up the multiplication square on page 44.
Complete these parts of it.

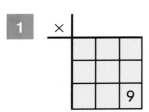 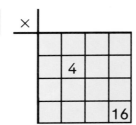 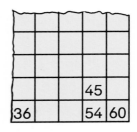

Uncover the multiplication square
to check your answers.

Let's investigate

These are parts of the multiplication square.
Multiply the corner numbers.

1	2
2	4

$1 \times 4 = \square$ $2 \times 2 = \square$

What do you notice?

Is it the same for these squares?
Is it the same for other squares?

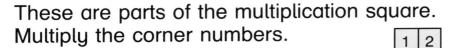

45

Area 3

A Find the area of these leaves.

1

beech

The area is ☐ cm².

2

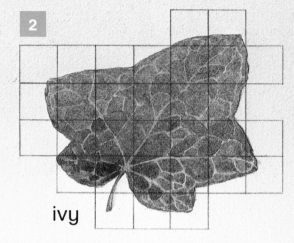

ivy

The area is ☐ cm².

3

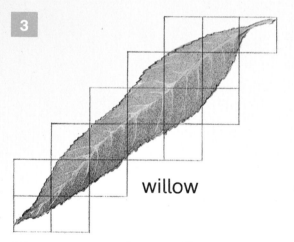

willow

The area is ☐ cm².

4

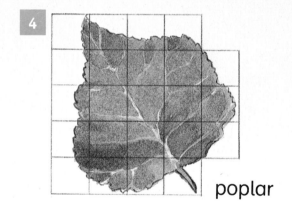

poplar

The area is ☐ cm².

Let's investigate

Draw two different leaves with the same area.

46

B   Find the area of these leaves.

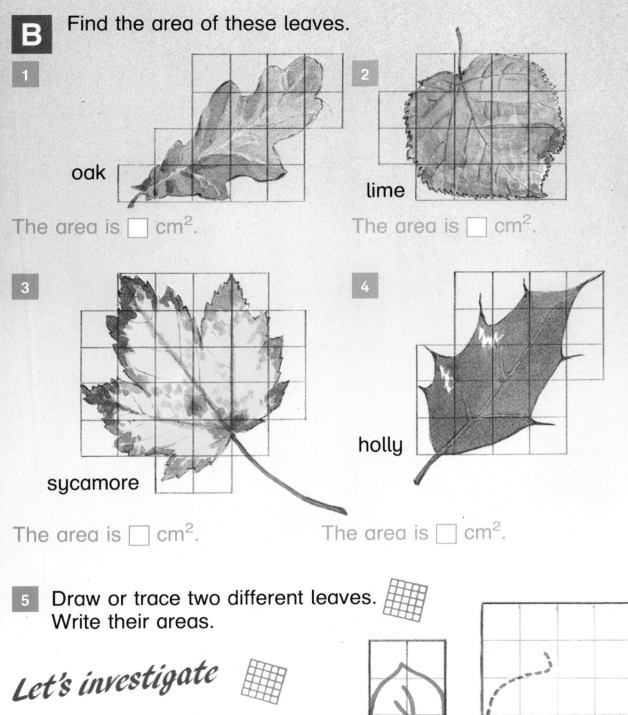

1 oak

The area is ☐ cm².

2 lime

The area is ☐ cm².

3 sycamore

The area is ☐ cm².

4 holly

The area is ☐ cm².

5   Draw or trace two different leaves.
Write their areas.

Let's investigate

Copy this leaf carefully.
Then draw it twice the size.
What do you notice about the two areas?
Is this the same for other leaf shapes?

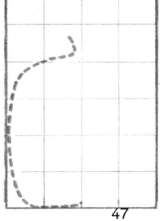

C

1 Use plain paper.
Fold it in half.
Draw half a leaf.
Cut it out.
Do not cut along the fold.
Estimate the area of the leaf.
Draw round the leaf on
squared paper.
Write its area.

2 What is the difference between
your estimate and the area of the leaf?

3 Try it again with a different leaf.

4 Was your estimate closer this time?

Let's investigate

Draw this on squared paper.
The total area of the leaves should be about 45 cm^2.

Number 14

A

1 24 people sit in double seats.
How many seats do they need?

2 2)48   **3** 2)52   **4** 2)74

5 242 people sit in double seats.
How many seats do they need?

2)242

6 2)268   **7** 2)484

8 2)660

9 There are 428 people on
the train.
How many double seats do
they need?

10 2)434   **11** 2)258

12 2)492   **13** 2)634

14 2)328   **15** 2)504

16 Some seats hold 3 people.
How many seats do 363 people need?

17 3)339   **18** 3)639   **19** 3)342

20 3)648   **21** 3)456   **22** 3)519

Let's investigate

Find numbers that will divide by both 2 and 3.

$$36 \text{ divides by 2 and 3.} \qquad 2)\overline{36} = 18 \qquad 3)\overline{36} = 12$$

B   **1**   There are 316 people on a train.
They sit in pairs. They need ☐ seats.

2 584 people are on another train.
Groups of four sit round tables.
How many tables do they need?

3 616 people sit in groups of four.
How many groups are there?

4 There are 56 tables on a train.
Four people sit round each table.
How many people are there?

5 Another train has 94 tables.
Four people sit round each table.
How many people are there?

6 4)652   **7** 7)917   **8** 6)684

9 4)272   **10** 3)825   **11** 5)465

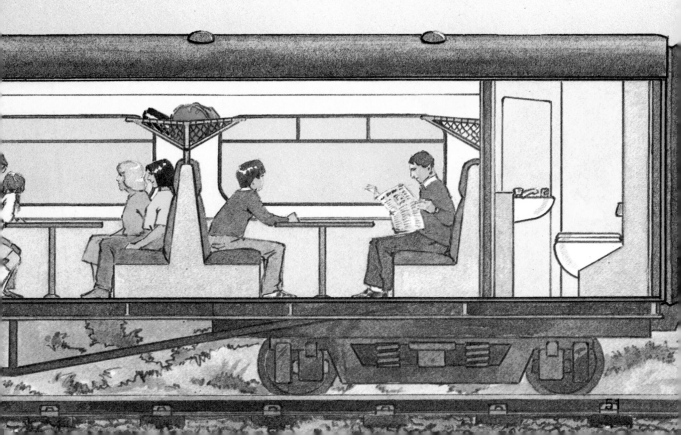

Time yourself

12 Copy this puzzle.
How long does it take
you to finish it?

Across
1 Divide 994 by 7.
3 Divide six hundred and
   seventy two by 3.
4 Share 928 between 4.
7 Ninety-six divided by 2.

Down
1 854 ÷ 7.
2 What is the answer if
   you divide 886 by two?
5 771 divided by 3.
6 Seventy-two shared by 3.

Let's investigate

621 divides exactly by 9.
Check this with a calculator.

Add the digits. $6 + 2 + 1 = \boxed{}$
What do you notice?

351, 603 and 126 divide exactly by 9.
Add their digits to check.

Find other numbers that will divide exactly by 9.
Try larger numbers.

Put ×, ÷ or = in the boxes.

1
32 ☐ 4 ☐ 8
32 ☐ 8 ☐ 4
4 ☐ 8 ☐ 32
8 ☐ 4 ☐ 32

2
72 ☐ 8 ☐ 9
8 ☐ 9 ☐ 72
72 ☐ 9 ☐ 8
9 ☐ 8 ☐ 72

3
4 ☐ 20 ☐ 80
20 ☐ 4 ☐ 80
80 ☐ 4 ☐ 20
80 ☐ 20 ☐ 4

4
3 ☐ 50 ☐ 150
50 ☐ 3 ☐ 150
150 ☐ 3 ☐ 50
150 ☐ 50 ☐ 3

5 Which number sentences can you find another answer for? Write them.

6 This is a division cross-number puzzle. Make up the clues.

Across
1
3

Down
1
2

Let's investigate

Make up numbers and clues for this division cross-number puzzle.

Across
1
2

Down
1

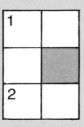

53

Data 3

A

Some children made a bar-line graph of their favourite colours.

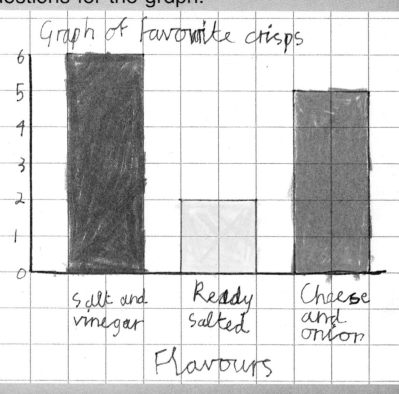

1 Which colour was the most popular?

2 How many liked red?

3 How many children were there altogether?

Some more children drew a bar chart of their favourite crisps.

4 Draw it as a bar-line graph.

5 Make up three questions for the graph.

6 Draw a bar-line graph for these buttons.

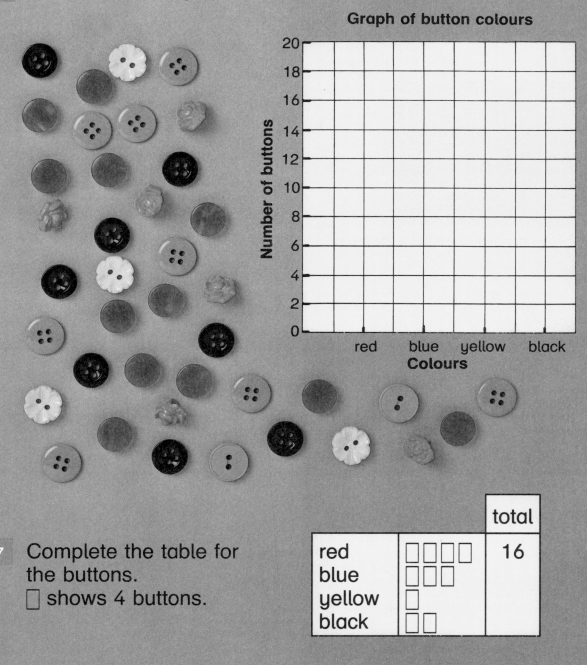

Graph of button colours

7 Complete the table for
the buttons.
☐ shows 4 buttons.

		total
red	☐☐☐☐	16
blue	☐☐☐	
yellow	☐	
black	☐☐	

Let's investigate

Look at the red buttons. Make a graph about them.
Make up some questions.

B

Here are the reaches of some children.

reach

Alan	140 cm
Barbara	150 cm
Carol	135 cm
David	120 cm
Emma	125 cm
Suna	130 cm

1 Measure your own reach.

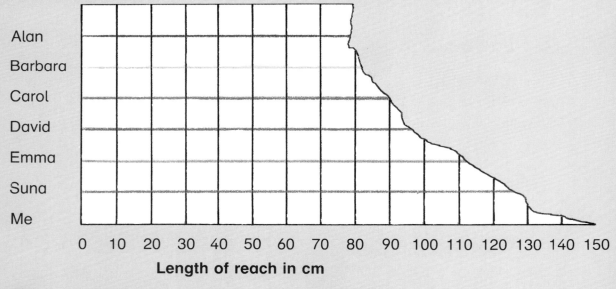

Length of reach in cm

2 Copy the bar-line graph and finish it.
Put your own reach on the graph.

3 What is Carol's reach?

4 Who has the longest reach?

5 Look at the graph.
Write the names in order of reach.
Put the shortest first.

Graphs of money

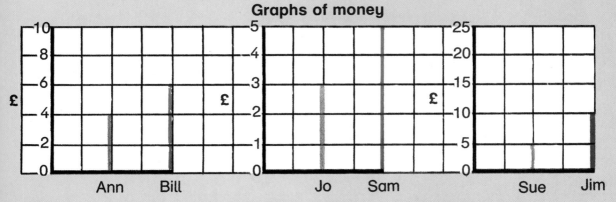

These bar-line graphs show amounts of money saved.

6 How much has each child saved?

7 Who has most money? **8** Who has least money?

9 Who has the same amount as Sue?

10 Who has twice as much as Jo?

11 These bar-line graphs all show the same amounts.
Draw them on squared paper.
Show the scale for each graph.

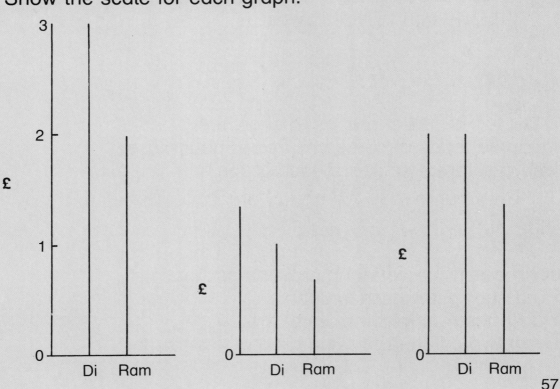

12 How much was collected each day?
Draw a bar-line graph to show this.

13 Which day was most money collected?

14 Which day was least money collected?
Why do you think this was?

Let's investigate

Draw two different graphs to show the
favourite foods of some children in your class.
Use a different scale for each graph.

C Let's investigate

John has twice as many crayons as Ann.
Paula has 2 fewer than John.
Harry has 4 more than John.
Draw two different bar-line graphs to show this.

58

Money 3

 1 Asad and Steven have 60p.
They share it equally.

They get ☐ p each.

Share these equally between two children.

2    **3**

4    **5**

6 90p ÷ 3   **7** 30p ÷ 3   **8** 36p ÷ 3

How much does each bag of sweets cost?

9 Jelly bears ☐ p   **10** Gums ☐ p

11 Mints ☐ p   **12** Toffees ☐ p

Jelly bears
2 for 48p

toffees
2 for 90p

mints
4 for 92p

gums
3 for 45p

Use coins.
Share £1·20 into two equal amounts.
Share £1·20 into three equal amounts.
Try sharing into other equal amounts.

B How much will these things cost?

1 one bag of chocolates.

2 a stick of rock.

3 one fudge bar.

4 one chocolate egg.

5 three sticks of rock.

6 3 chocolate eggs.

7 one cake.

8 2 chocolate eggs.

9 two bags of chocolates.

10 two sticks of rock.

rock
6 for 96p

chocolates
3 for £3·30

chocolate eggs
6 for £1·32

cakes
2 for £2·36

fudge
3 for £1·29

The boxes with yellow bows are sold at half price.
What would you pay for them?

11 ☐ p for the £1·50 box.　　**12** £ ☐ for the £4·50 box.

Let's investigate

Use coins. Which amount on the price chart
will divide by 2, 3, 4, 5 and 6?
Find different amounts which divide by all these numbers.

C　Look at the boxes with red ribbon.

1　The most expensive price for one box is £ ☐.

2　What is the cheapest price for one box?

Let's investigate

Make up some prices to complete the labels.
The difference between the cheapest and
most expensive price for one must be 5p.
Find different ways to do this.

3 for £ ☐

2 for £ ☐

£ ☐ each

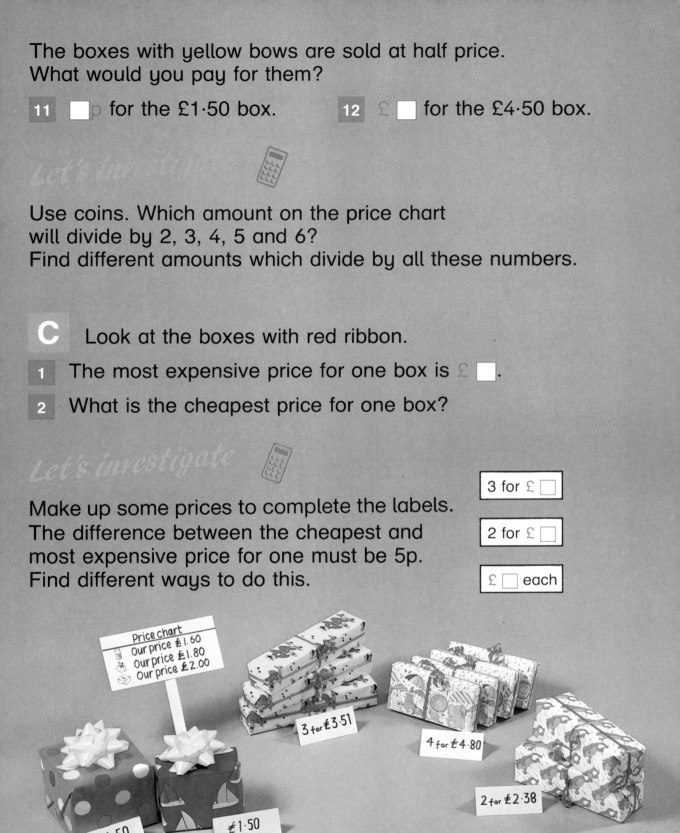

Price chart
Our price £1.60
Our price £1.80
Our price £2.00

3 for £3·51

4 for £4·80

2 for £2·38

£4·50

£1·50

Number 15

A

One pane of glass is broken.
This is $\frac{1}{8}$ of the whole window.

1 ☐ panes are not broken.

2 This is $\frac{\square}{8}$ of the window.

> In the days of Queen Anne a gentleman's house would often have a window above the front door. It was shaped like a fan and called a fanlight.

3 $\frac{3}{8}$ of the window is ☐ panes.

4 $\frac{5}{8}$ is ☐ panes.

5 6 panes is $\frac{\square}{8}$.

6 2 panes is $\frac{\square}{8}$.

7 Fold a circle into 8 equal parts. Write $\frac{1}{8}$ on each part.

8 Colour $\frac{1}{2}$ of the circle red. Colour $\frac{1}{4}$ blue. Colour $\frac{1}{8}$ green.
What fraction is left?

9 $\frac{1}{2} = \frac{\square}{8}$

10 $\frac{1}{4} = \frac{\square}{8}$

11 $1 = \frac{\square}{8}$

Let's investigate

Fold a square into $\frac{1}{8}$s.
How many different ways can you do it?

B Sash windows were first seen in England in the 17th century. They open by sliding up and down on cords.

1 There are ☐ panes in $\frac{1}{2}$ of the whole window.

How many are in $\frac{4}{8}$ of it?

2 There are ☐ panes in $\frac{1}{4}$ of the whole window.

How many are in $\frac{2}{8}$ of it?

Use the fraction wall to help you find the missing fractions.

3 $1 = \dfrac{\square}{2}$
4 $1 = \dfrac{\square}{4}$
5 $1 = \dfrac{\square}{8}$
6 $\dfrac{1}{2} = \dfrac{\square}{4}$

7 $\dfrac{1}{2} = \dfrac{\square}{8}$
8 $\dfrac{1}{4} = \dfrac{\square}{8}$
9 $\dfrac{2}{4} = \dfrac{\square}{8}$
10 $\dfrac{3}{4} = \dfrac{\square}{8}$

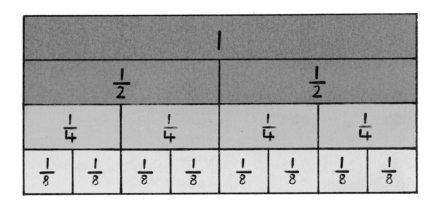

11 How many panes of glass are in half the window?

12 How many are in quarter of it?

13 $\frac{1}{8}$ of 16 is ☐.   **14** $\frac{3}{8}$ of 16 is ☐.

15 $\frac{7}{8}$ of 16 is ☐.   **16** $\frac{5}{8}$ of 16 is ☐.

Let's investigate

Draw a window with 24 panes.
Colour $\frac{1}{2}$ red, $\frac{3}{8}$ blue, $\frac{1}{8}$ green.
Write the fractions on.
Find other ways to do it.

C

1 $\frac{1}{2}$ of this window is ☐ panes.

2 $\frac{3}{4}$ of this window is ☐ panes.

3 $\frac{1}{8}$ of this window is ☐ panes.

4 What fraction of it is 15 panes?

Let's investigate

Put numbers in the boxes to make different fraction sentences for 40.

$\frac{☐}{8}$ of 40 is ☐.

What do you notice about each pair of numbers that you write?

Do the same for $\frac{☐}{8}$ of 32 is ☐.   What do you notice this time?

Length 3

Zoo information

A

◄ Giraffe
599 cm high

Black headed heron ►
95 cm high

Emperor penguin
115 cm high

◄ Gorilla
172 cm high

Honey badger
25 cm high ▼

Ostrich ►
230 cm high

100 cm = 1 metre or 1 m

130 cm = 1 m 30 cm or 1·30 m

65 cm = 0·65 m

1 The penguin is ☐ · ☐ m.

2 The giraffe is ☐ · ☐ m.

3 The gorilla is ☐ · ☐ m.

4 The heron is ☐ · ☐ m.

5 The ostrich is ☐ · ☐ m.

6 The badger is ☐ · ☐ m.

7 How tall are you? ☐ cm = ☐ · ☐ m.

8 Which of the animals are taller than you?

9
```
  m
  4·27
+ 2·13
_____
```

10
```
  m
  2·50
+ 3·64
_____
```

11
```
  m
  5·63
+ 1·84
_____
```

12
```
  m
  3·29
+ 4·53
_____
```

65

The tawny owl

Wingspans	
Albatross	3·05 m
Golden eagle	2·20 m
Tawny owl	1·00 m
Robin	0·22 m

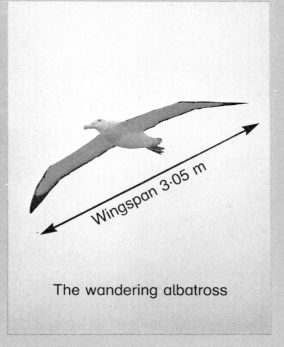

Wingspan 3·05 m

The wandering albatross

13 Fill in this chart.

Albatross	wingspan ☐ cm
Golden eagle	wingspan ☐ cm
Tawny owl	wingspan ☐ cm
Robin	wingspan ☐ cm

14 How wide is your reach?

15 How much wider is it than the wingspan of the owl?

16 m	**17** m	**18** m	**19** m
5·84	6·71	9·86	7·26
− 2·37	− 4·35	− 3·18	− 1·60

Let's investigate

Measure the reach of some of your friends.
Which ones add up to about the same length as
the wingspan of the albatross?

Marine information

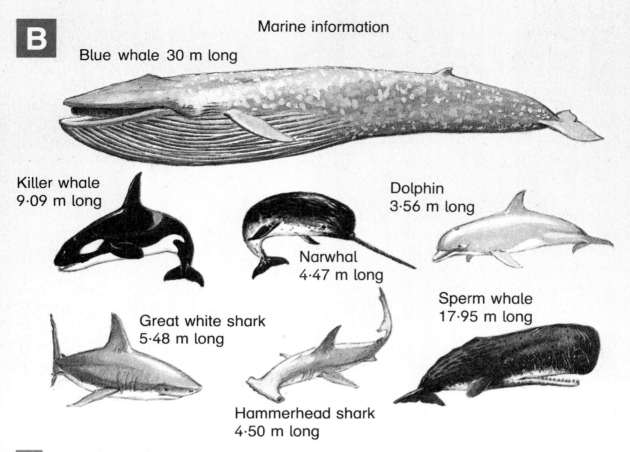

Blue whale 30 m long

Killer whale
9·09 m long

Dolphin
3·56 m long

Narwhal
4·47 m long

Sperm whale
17·95 m long

Great white shark
5·48 m long

Hammerhead shark
4·50 m long

1 How much longer is the killer whale than the dolphin?

2 What is the total length of the narwhal and the killer whale?

3 How many cm longer than the hammerhead shark is the great white shark?

4 Find the total length of the two sharks.

5 Find the difference in length between the dolphin and the longer shark.

6 Measure the length of your classroom. Would a sperm whale fit into it?

7 Would a blue whale fit into your playground?

8 How many buses would fit along the blue whale?

A bus is about 10 m long.

9 Write the lengths
of these eggs.

	cm	m
ostrich	17	☐
hen	6	☐
mute swan	☐	0·11
blackbird	3	☐

> The male ostrich digs a nesting hole with his beak. One or more females lay their eggs in it. There are usually more than 30 eggs in the nest.

10 Draw two of the eggs. Measure them carefully.

Let's investigate

Draw an animal.
Make the tail twice as long as the body.
Make the legs half as long as the body.
Write the measurements.

C A champion American flea
jumped nearly 20 cm.
This was 130 times its
own height.

1 The tallest giraffe was about 6 m high.
How high would it have jumped
if it had jumped as well as the flea?

Let's investigate

Find your height to the nearest metre.
How high could you jump if you were like the flea?
What can you find out about your height and your best
high jump?

Weight 3

The parcel weighs 1500 g.
The scale shows 1 kg 500 g.

1 kg = 1000 g
500 g = 500 g
1500 g = 1 kg 500 g or 1·500 kg.

Write the weights in kilograms.

1 1400 g   **2** 1250 g   **3** 2250 g   **4** 1640 g

☐ . ☐☐☐ kg

5 Add the two heaviest weights.

6 Add the two lightest weights.

kg
☐ · ☐☐☐
+ ☐ · ☐☐☐
_____

7 What is the difference in weight between the heaviest and the lightest parcel?

69

8 kg	**9** kg	**10** kg	**11** kg
2·702	3·540	4·672	1·568
+ 1·189	+ 1·270	− 1·351	− 1·134
———	———	———	———

Let's investigate

Make some parcels. Find things to put in them.
Make parcels weighing 2 kg and 4 kg.
Make a list of the things in each parcel.

B

What is the weight in each bag?

5 Which is the heaviest bag?

6 Which is the lightest bag?

7 10 apples weigh 1·350 kg.
9 apples weigh 1·200 kg.
The missing one weighs ☐ kg.

8 5 bananas weigh 0·760 kg.
4 bananas weigh 0·610 kg.
The missing one weighs ☐ kg.

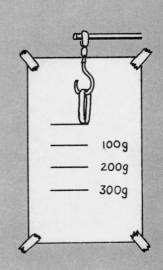

Let's investigate

Find things in your classroom that have a weight
difference of about 100 g, 200 g, or 500 g.

C **1** How could you weigh a cat that refuses
to sit on the scales?

Let's investigate

Hang a rubber band on a hook.
Stick a piece of paper behind it.
Mark where the band hangs to.

Hang 100 g, 200 g and 300 g weights
on the band and mark how far it
stretches each time.
Try other objects and work out their weights.

100g
200g
300g

Volume

These shapes are built with centimetre cubes.
When more cubes are used the volume is greater.

1 Make these shapes.
Which shape has the greater volume?

2 How do you know?

3 Use 8 centimetre cubes each time,
Build three different shapes.
What is the volume of each shape?

Let's investigate

Use 12 cubes each time.
How many different cuboids
can you make with them?

How many different cuboids can you make
with 15 cubes?

B How many cubes will fill the holes?
The holes go right through the shapes.

1

2

How many cubes will fill these boxes?

3

4

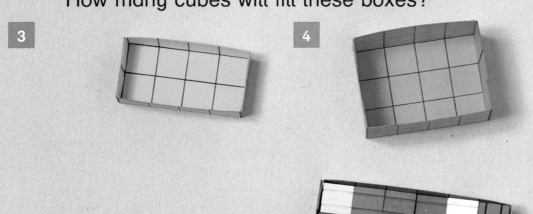

5 How many more cubes
are needed to fill
this box?

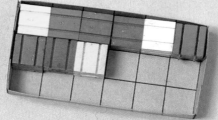

6 Draw this shape.
Use 24 cubes to
build a cuboid on it.
How many layers are there?

Let's investigate

Draw some more shapes on squared paper so that
you can build a cuboid with 24 cubes on each one.

C

1 Make a box to hold 16 centimetre cubes exactly.

2 Use cm cubes to
make a shape
which has a hole
the size of 1 cube.

What is the volume
of the shape?
What is the volume
of the hole?

3 Do the same for a shape which has
a hole the size of 2 cubes.

Let's investigate

Build some shapes with holes larger than 2 cubes.
Make a chart showing the volumes of the shapes and
the volumes of the holes. What do you notice?

Time 3

A

1. What day is 10th December?

2. How many Tuesdays in December?

3. What is the date on New Year's Day?

How many days are there in these months?

4. June
5. March
6. April
7. October

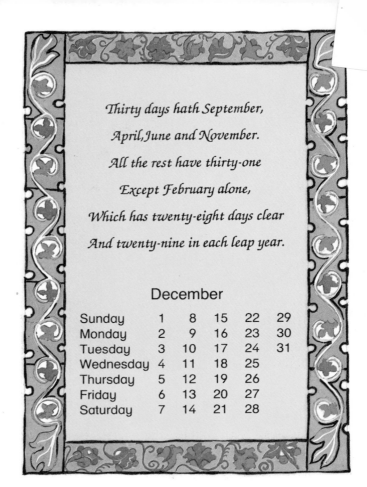

Thirty days hath September,

April, June and November.

All the rest have thirty-one

Except February alone,

Which has twenty-eight days clear

And twenty-nine in each leap year.

December

Sunday	1	8	15	22	29
Monday	2	9	16	23	30
Tuesday	3	10	17	24	31
Wednesday	4	11	18	25	
Thursday	5	12	19	26	
Friday	6	13	20	27	
Saturday	7	14	21	28	

8. Make a month clock like this. Use it to help you write these dates another way.

21 July 1991	21.7.91	
9. 25 June 1992		
10.	15.12.95	
11. 3 April 1996		
12. 18 May 1995		
13.	29.8.93	

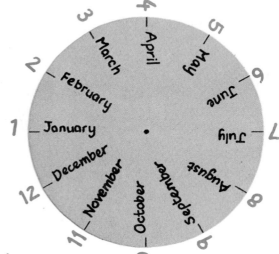

Let's investigate

Write six dates which are important to you.
Say why you have chosen them.

Opening Times
(10 am to 6 pm)
January to March
   Saturdays only
April to June
   Monday, Tuesday
   Friday, Saturday
July to September
   daily
October to December
   Saturday, Sunday

1 On which day is the castle open every month of the year?

2 For how many months is the castle open on a Monday?

3 In which months is the castle open on a Sunday?

4 Is the castle open on Friday 3 May?

				1991		
S	M	T	W	T	F	S
					1	2
3	4	5	6	7	8	9
10	11	12	13	14	15	16
17	18	19	20	21	22	23
24	25	26	27	28		

				1991		
S	M	T	W	T	F	S
					1	2
3	4	5	6	7	8	9
10	11	12	13	14	15	16
17	18	19	20	21	22	23
24	25	26	27	28	29	30
31						

				1991		
S	M	T	W	T	F	S
	1	2	3	4	5	6
7	8	9	10	11	12	13
14	15	16	17	18	19	20
21	22	23	24	25	26	27
28	29	30				

These months follow each other.

5 Which months are they?   6 Is it a leap year?

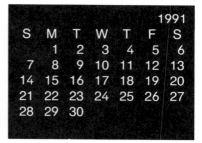

7 Draw a calendar for the month after the three shown.

8 The date 4.9.91 is a Wednesday.
   Draw a calendar for the month.

This is part of a register.

9 Who is the oldest?

10 Who is the youngest?

11 Whose birthday is in May?

NAME	DATE OF BIRTH		
	D	M	Y
Jane Andrews	21	1	83
Tarun Ashar	3	6	83
Ajit Kaur Bassi	14	2	83
Lucy Dixon	25	5	83
Emma Ford	9	3	83

Let's investigate

Write the names and dates of birth of some children
in your class. Put them in order.
Make up questions about them.

C

Use a year calendar.

1 What day is the
   50th day of the year?

2 What day is the
   100th day of the year?

3 What day is the
   300th day of the year?

4 How many days of the
   year have passed on 1 April?

Let's investigate

Make up some questions for a
friend about the year calendar.
Check their answers.

Angles 3

In 1800 fans were quite large and often had pictures on them. Handles were sometimes made of ivory. Some fans were made of feathers.

1 A right-angle is ☐°.

2 This is a straight angle. It is ☐°.

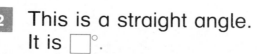

An acute angle is smaller than 90°.

An obtuse angle is between 90° and 180°.

Use a right-angle and write the names of these angles.

3

4

5

6

7

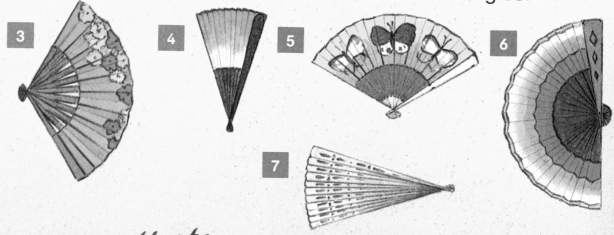

Let's investigate

Draw some different fans that show acute, obtuse and straight angles. Label them.

78

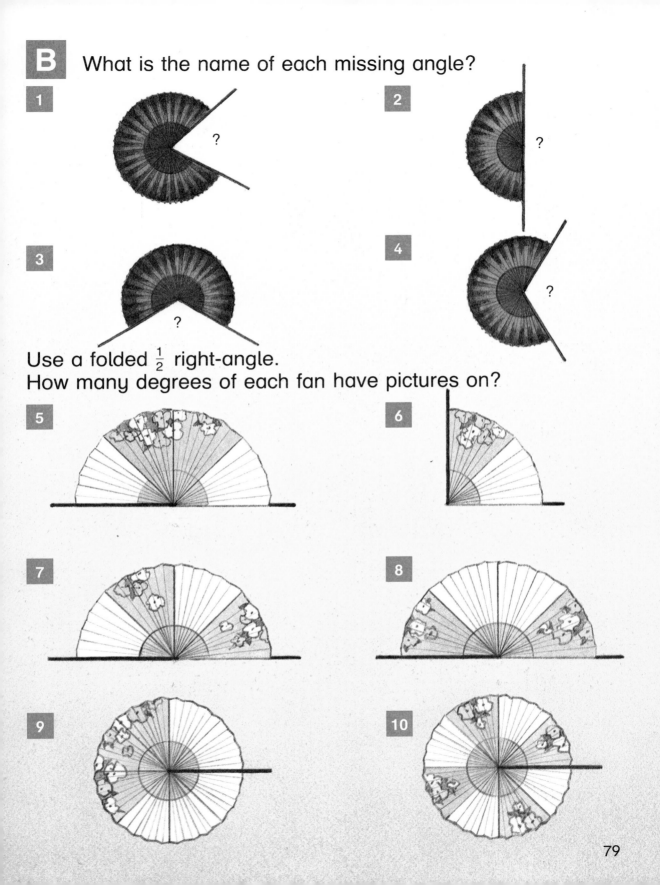

B What is the name of each missing angle?

1

?

2

?

3

?

4

?

Use a folded ½ right-angle.
How many degrees of each fan have pictures on?

5

6

7

8

9

10

Let's investigate

Draw a fan. Colour it.

90° of it must be red.
90° of it must be green.
45° of it must be yellow.

Find different ways to do it.

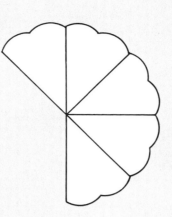

C

1 Draw a fan that is opened to 180°.
Divide it into 45° parts and colour it.

Let's investigate

Fold a 45° angle.
Use it to draw different angles.
What size angles can you make?